I0468023

NUREG-1148

Nuclear Power Plant Fire Protection Research Program

Manuscript Completed: April 1985
Date Published: July 1985

A. Datta

Division of Engineering Technology
Office of Nuclear Regulatory Research
U.S. Nuclear Regulatory Commission
Washington, D.C. 20555

Foreword

This document presents a plan for Nuclear Power Plant Fire Protection Research to be performed principally by the Electrical Engineering Instrumentation and Control Branch, Division of Engineering Technology, Office of Nuclear Regulatory Research. It presents the objectives of the program, describes the research that will be carried out to achieve the objectives, and identifies areas of coordination with the Office of Nuclear Reactor Regulation.

We perceive this plan to be a living document and expect to revise it periodically to take into account our experience in implementing the plan and comments received from interested parties within the NRC and among the public.

Comments on this document are welcome at any time and will be considered in the development of subsequent editions of this plan. They need not be restricted to the research activities described herein; comments identifying omissions or recommending additional research are also welcome.

Bill M. Morris, Chief
Electrical Engineering Instrumentation
and Control Branch

Approved by:

Guy A. Arlotto, Director
Division of Engineering Technology
Office of Nuclear Regulatory Research

TABLE OF CONTENTS

Page

I. INTRODUCTION... 1

    1.1 Program Overview...................................... 1

    1.2 Background.. 2

         1.2.1 Contribution of Fire to Risk 2
         1.2.2 Control Room.................................. 4
         1.2.3 Actuation of Fire Suppression Systems.......... 5

II. APPROACH... 7

    2.1 Program Organization.................................. 7
    2.2 Fire Source Characterization.......................... 7
    2.3 Fire Environment Determination........................ 8
    2.4 Fire Failure Thresholds of Equipment.................. 9
    2.5 Control Room Habitability and Fire Safety Study (CR)... 10
    2.6 Coordination with RMIEP 12

III. PROJECTED ACCOMPLISHMENTS, SCHEDULES, AND BUDGET.......... 13

APPENDIX A.   DETAILED PROJECT DESCRIPTIONS.................... 17

    A.1 Fire Source Characteristics (FSC)..................... 17
    A.2 Fire Environment Determination (FED).................. 17
    A.3 Fire Failure Thresholds (FFT)......................... 19
    A.4 Control Room Habitability and Fire Safety Study (CR)... 20

APPENDIX B.   COORDINATION WITH NRR/CMEB...................... 22

    B.1 Fire Source Characteristics (FSC)..................... 22
    B.2 Fire Environment Determination (FED).................. 22
    B.3 Fire Failure Thresholds (FFT)......................... 23

FIRE PROTECTION RESEARCH PROGRAM

I. INTRODUCTION

1.1 Program Overview

The goal of the Fire Protection Research Program is to develop test data and analytical capabilities to support the evaluation of:

1.   the contribution of fires to the risk from nuclear power plants,

2.   the effects of fires on control room equipment and operations, and

3.   the effects of actuation of fire suppression systems on safety equipment.

These three goals will be reached by implementing a common threefold research approach:

Define Fire Sources. A range of fire sources will be characterized with respect to their energy and mass evolution, including smoke, corrosion products, and electrically conductive products of combustion. The combustible content and configurations of the sources determine the energy and mass release rate characteristics of the fires.

Define Environments. An analytical method for determining the environment resulting from fire will be developed. This method will account for the source characteristics, the suppression action following detection of the fire, and certain parameters specific to the plant enclosure in which the fire originates, such as the geometry of the enclosure and the ventilation rate. The developing local environment in the vicinity of safety-related equipment will be expressed in terms of temperatures, temperature rise rates, heat fluxes, and moisture and certain species content.

<u>Define Equipment Response</u>. The response of certain safe shutdown equipment and components to the environmental conditions will be studied. The objective will be to determine the limits of environmental conditions that a component may be exposed to without impairment of its ability to function.

The rationale for this approach is provided in Section II, and the detailed project descriptions are in Appendix A. The schedules for completing the separate portions of the fire protection research program and the budget estimates are shown in Tables 1 and 2.

<u>Coordination With Other Programs</u>. Development and evaluation of this Fire Protection Research Program is being coordinated with the work in progress on Control Room Habitability Program and the Risk Methods Integration and Evaluation Program (RMIEP). The latter will be evaluating all risks associated with the La Salle plant.

1.2 <u>Background</u>

1.2.1 <u>Contribution of Fire to Risk</u>

NRR research needs in PRA methodology were outlined in a memo from Denton to Minogue dated November 30, 1982. The recommendations in the area of external event risks included information to be generated on "equipment performance in accident environment," "suitable but simple methods ... to predict variations in environmental parameters following an accident, for example: temperature and pressure build-up," and "fragility data relating equipment failure probabilities to environmental changes." Specifically, in the area of fire risk, RES was urged to give attention to the aspects of "systems interaction between fire protection features and safety systems," "the reliability of fire protection features" and "the likelihood of qualified equipment to withstand the effects of fire and fire suppressants."

The External Events PRA Working Group, in its recommendations for improving the capability of PRA for external events ("External Events Considerations in Probabilistic Risk Assessment," dated June 30, 1983), observes that "with fire, as with all the other disciplines, the insufficiency of the data base is a major problem. In addition, the validity of modeling techniques, the knowledge of fire accident sequences, and the knowledge of the reliability of the fire protection systems limits the degree of confidence in PRA results."

The working group has specifically identified the following information needs:

1. The frequency/magnitude of fires in power plants, including fires in transient materials;

2. The distribution of "time-to-fire detection" as it relates to the location and the magnitude of plant fires;

3. The distribution of "time-to-fire suppression" conditional upon the location and the magnitude of plant fires;

4. Component responses and their damageability from fires of different magnitudes;

5. Failure rate data on fire protection systems;

6. The frequency of fires simultaneous with secondary independent initiating events.

Both NRR and RES have initiatives underway to address most of these needs. The Chemical Engineering Branch of NRR has awarded a contract to the Brookhaven National Laboratory to establish a methodology and a data base for probabilistic assessment of fire risks and to develop guidance for NRR evaluation of PRAs by others.

The Division of Risk Analysis and Operations of RES is engaged in the Risk Methods Integration and Evaluation Program (RMIEP). The program has as its objective the development of an improved PRA method and demonstration of the method in reference to one actual plant, the La Salle County Nuclear Station, Unit 2. The Fire Risk Analysis part of the RMIEP proposes to reduce the uncertainties in the PRA models for ignition, fire growth, detection and suppression, and component fragility. PRA methods currently rely on past experience with fire incidents in large industrial plants, both nuclear and non-nuclear. There are serious weaknesses in that data base in the areas of fire growth and consequent threat to barriers and safety-related equipment, detection and suppression times, and component damage information. The program described here will complement the RES/DRAO program by generating deterministic information in these areas, both by experimentation and by analysis. The two programs (those of DET and DRAO of RES) are coordinated so that the La Salle fire PRA when completed will have input from an improved data base.

Future applications of PRA methods are foreseen in the assessment of fire risks arising from secondary independent initiating events, such as earthquakes. The data base on equipment fragility and the analysis methodology for fire safety margins proposed to be developed in this program would be useful for that eventual application.

1.2.2 Control Room

In the control room, there is a high density of electrical instrumentation and control cables of the redundant trains in necessarily close proximity to each other. Fire protection of the control room is complicated by the fact that malfunction of critical components may have cascading effects of spreading loss of control in areas remote from the control room. The requirements of operability for components in the electrical cabinets in the control room, therefore, become more stringent. A systematic study of the control system and how faults may propagate in the event of a fire is needed. A systems approach, utilizing the knowledge and the methodologies of spatial separation, should be used in such a study. The data base produced would be of value in devising ways of ensuring a margin of safety.

The other concern in the control room, as expressed by the ACRS on several occasions, is that existing requirements to protect the occupants of the control room in accident situations may not be adequate (ACRS to the Chairman, NRC, dated March 11, 1980). NRR has responded to this concern (Denton to Dircks, dated June 29, 1984) with a program plan contained in the staff report titled Control Room Habitability, dated June, 1984. One of the recommendations of the staff report is that "limiting environmental conditions for operation

in the control room should be established and should consider human performance
as well as equipment operation as the basis for selection of appropriate limits."
The staff report itself proposes environmental criteria for human performance.
The need is for a determination whether the generic control room habitability
system is adequate for the maintenance of that minimal environment through a
credible fire accident. The effectiveness with which the habitability system,
which includes the HVAC system, isolates the control room and removes smoke and
toxic products from within needs to be studied, and possible remedial actions
need to be examined. This program addresses the issue by determining the
fire-generated environment and the purge effectiveness.

1.2.3 Actuation of Fire Suppression Systems

The Office of Inspection and Enforcement describes in Information Notice 83-41,
which was addressed to all holders of operating licenses and construction per-
mits, several instances of automatic fire suppression system actuations resulting in inoperability of safety-related equipment. The document provides a list
of twelve "selected examples" of reported events occurring between November 1980
and October 1982. In some of these examples, the suppression system actuated
properly in response to a valid signal. In others, there was no real need for
actuation. The latter cases do not appear to have any single causative factor.
Errors apparently have been made in design, installation, or operating and maintenance procedures.

In one incident an ionization-type smoke detector initiated an alarm because
of a buildup of humidity and steam vapor in the HPCI room. The consequent
automatic actuation of the water deluge system rendered the HPCI system inoper-
able. In another incident, inadvertent actuation of the suppression system
resulted in the tripping of an RPS motor-generator set and in a small amount
of water entering the control rod drive switchgear cabinet. The reactor was
manually tripped from full power in response to two dropped control rods
following the condition.

Furthermore, concerns have been expressed (Michelson to Denton, dated July 28,
1982) about the degradation in the performance of equipment, including solid

state devices, due to the cooling effect of spuriously released CO_2 fire suppressant.

Any operation of an automatic fire suppression system will create environments that may damage safe shutdown equipment. In addition to a general lack of information concerning the effects of fire on safe shutdown equipment, we also know little about the effects of fire suppressants on them. Current regulations for equipment qualification require that, only the equipment that would be exposed to high-energy steam line breaks classified as DBAs must be qualified for steam and humidity. The need exists to determine the effects of fire suppression agents, in addition to the effects of fire, on the operability and failure thresholds of safe shutdown equipment.

II. APPROACH

2.1 Program Organization

The threefold approach outlined in Section 1.1, above, constitute the first three segments of the Fire Protection Research Program, wherein the basic methodology and data are developed. Work on each of these segments will initially proceed simultaneously and independently, but as the program matures results of the segments will be combined. For example, use of data from the Fire Source Characterization (FSC) in the methodology developed by the Fire Environment Determination (FED) will yield the progressively developing fire environment in a given plant enclosure. That fire environment used in conjunction with the results of the Fire Failure Thresholds (FFT) testing will enable determination of the time-to-failure of particular safe shutdown components located in the plant enclosure of interest. This quantity (the time-to-failure) is one of two key ingredients in the eventual determination of component fragility. The other is the time-to-suppression. The probability of the suppression time exceeding the failure time is the starting point of fragility analysis in PRA. The three segments (FSC, FED and FFT) all produce data and methodologies that are generic for the rest of the Fire Protection Research Program. That includes the evaluation of the fire safety margin for control rooms, the interaction of the fire suppression system with safe shutdown system, and the generation of a data base for PRA. A rational for each of these segments of the program, its plan of operation, and its expected end products follow.

2.2 Fire Source Characterization (FSC)

Fires in nuclear power plants initiate and grow in a finite number of combustibles in known configurations (e.g., vertical and horizontal cable trays) and unknown configurations (e.g., lubricants, solvents, and trash). The energy and mass release rates of the fire, including release rates of certain toxic and corrosive products of combustion, need to be known in order to determine the environments resulting from the fire.

The end product of FSC will be the characteristics of a range of fires that could occur in nuclear power plants. At this time, the set is expected to include typical cable system fires and fires resulting from transient solid combustibles and flammable liquid spills. Some experimental data on the heat and the mass release rates of such fires exist in the literature. These data will be pooled and the gaps in the information will be filled by further experimentation. Fire characteristics will be determined for classes of fuel, and for particular configurations of the fuel, in the form of heat and the mass release rate histories.

Appendix A, Section A.1, contains a discussion of program details.

2.3 Fire Environment Determination (FED)

A series of reproducible, full-scale tests will be performed in which a range of fires studied under FSC (Section 2.2) above will be replicated, and a thermal mapping in three dimensions will be obtained by actual measurement of such parameters as temperature, heat flux, and species concentration. Since it is impossible to replicate all combinations of nuclear power plant enclosures and other variables such as door opening, ventilation rates, placement of the fire, and the fire characteristics, this test series will cover a range of enclosure-variables, some with representative equipment configurations.

An analytical capability will be developed to calculate potential fire environments within enclosures, given the fire source characteristics. An existing basic computer code will be adapted to the specific needs of this program and will be validated by the test series.

The model development and testing work will contribute to two related areas. First, the technique of analysis using the computer code can be used to examine various configurations of barriers, shields, and enclosures to determine whether they improve on the margin of safety of the safe shutdown capability. Second, with the analysis of the flow-field for a given configuration of equipment in an enclosure, one obtains a sound technical basis for the siting of fire detectors and fire suppression equipment.

The end products of FED will be:

1. A data base of test results showing fire environment developed for a
    range of potential fires in a set of plant enclosures covering a range of
    dimensions.

2. A user-interactive computer program applicable to a wide range of geometries
    for the purpose of predicting fire environments in enclosures.

3. An integrated methodology whereby the environments created by a range of
    fires (FSC, Section 2.2) in plant enclosures is determined, and in conjunc-
    tion with equipment damage thresholds (FFT, Section 2.4) is used to estimate
    the time-to-failure of certain classes of equipment.

4. A data base for development of fire PRA in nuclear power plant enclosures,
    including time-to-detection, time-to-suppression, and the probable sequence
    of failure of equipment in potential fire environments.

Appendix A, Section A.2 contains a discussion of program details.

2.4 Fire Failure Thresholds of Equipment (FFT)

The level of degradation due to fire environments at which the continued operability of selected safe shutdown equipment can no longer be ensured will, for
the purpose of this program, be designated the fire failure threshold of the
equipment. The fire environment includes the contribution of any extinguish-
ment action and the actuation of an automatic fire suppression system without a
fire being present. Although it is common to find operating temperature ratings
on electrical equipment, fire failure threshold ratings are not specified. Fire
failure thresholds of equipment will be related to parameters of the environment
in the immediate vicinity of the equipment. These parameters will, as a minimum,
include temperature and heat flux. Other parameters, such as corrosive chemical
species concentration and smoke density, may also be pertinent to the loss of
operability of specific equipment. A fire failure, furthermore, may be due to
a combination of the environmental parameter and the duration of exposure. The
fire failure will be a loss of function, e.g., a motor fails to start up because

of damage to a cable, or due to malfunction or spurious actuation, e.g., a relay spuriously makes or breaks a circuit.

Surveys to identify the present state of knowledge of the operability requirements of specific equipment and the probable mechanisms of degradation of that equipment in fire environments will be made. These surveys will draw upon PRAs, FSAR licensee submittals, and NRR and I&E correspondence. Another source of information is the plant design and operational experience of architect/engineer firms.

A test program to augment the available data on the fire failure thresholds will follow the surveys. This test program will employ an apparatus capable of reproducing simulated fire environments, such as increasing and decreasing temperature ramps associated with the development of actual fires, automatic fire suppression activities, and selected combustion product concentrations. The apparatus will also be capable of determining the instant of component or system failure under the test conditions. The program will include tests to study degradation of selected equipment due to inadvertent application of fire suppressants, such as water and CO_2.

The end products of FFT will be:

1. A data base on the fire and fire suppressant failure thresholds of selected safe shutdown equipment, and

2. Methodologies for the determination of fire failure thresholds of such equipment.

Appendix A, Section A.3 contains a discussion of program details.

2.5 Control Room Habitability and Fire Safety Study (CR)

The study of control room habitability and fire safety will combine the methodologies of fire source characterization, determination of the resulting environment, and safety-related equipment response. The equipment of concern will include those related to both safe shutdown and to the maintenance of habitability.

Transient fire sources in the control room will be characterized in FSC (Section 2.2). These fire sources will be used to ignite certain instiu combustibles to determine the range of energy, smoke, and toxic gas output of fires in which one or more of the electrical control cabinets are involved in fire. This set of reference fire data will be used to determine the fire environment and smoke removal needs in generic situations and to validate the analysis technique.

An analysis using a three-dimensional hydrodynamic computer code (Section 2.3) will map the air and combustion product flow under various arrangements of the ventilation and smoke purge system. Given the criteria of minimal human habitability from the NRR Division of Human Factors Safety, this analysis will estimate the margin of time an operator has in a given scenario to initiate fire suppression action, or to transfer control to a remote shutdown panel and vacate the control room.

A test program will replicate control room configurations with postulated fire sources and will record the developing environment in each case. Equipment fire failure thresholds will be determined in FFT (Section 2.4). The hazardous environment quantified by the test program will be used to estimate the extent of equipment damage or malfunction in the control room. The extent of propagation of the electrical malfunction that may cause loss of plant shutdown capability will be studied in a control system analysis.

The end products of the Control Room Habitability and Fire Safety Study will be:

1. A data base on a range of fire scenarios that could cause failure of equipment and lead to conditions resulting in loss of shutdown and safety functions;

2. The system interactions affecting shutdown and safety functions for the identified failures; and

3. Estimate of the smoke removal effectiveness of typical habitability systems.

Appendix A, Section A.4 contains a discussion of program details.

2.6 Coordination with RMIEP

The RMIEP Fire Risk Analysis program calls for deterministic data in several
areas to reduce uncertainties in the various models used in the fire portion
of the PRA. The RMIEP is to demonstrate the improved methodology by performing
a PRA for the La Salle plant using an analysis of the fire growth and spread
probabilities for key single and multiple fire areas. The fire environment computer code, proposed to be developed as part of this program (Section 2.3),
will be used to estimate the environments created by potential fires in some of
the most complicated of those areas, such as the control room.

The RMIEP calls for deterministic and probabilistic analysis of fire detection
and suppression times and component fragility. The full-scale fire tests and
the computer code of this program (Section 2.3 and 2.5) will generate estimates
of the time-to-detection and the time-to-suppression of fire. The fire failure
thresholds (Section 2.4) of safety-related equipment and cables created by this
program will be used to reduce PRA uncertainties in component fragility.

III. PROJECTED ACCOMPLISHMENTS, SCHEDULES, AND BUDGET

The projection of accomplishments from the Fire Protection Research Program efforts presented below is based on the level of funding shown in the proposed budget figures through FY 1986. Table 1 shows the projected time schedule and Table 2 the budget estimates. The parenthetical references are to projects in Appendix A, below.

FY 1984 Accomplishments

1. Survey of fire characteristics from past test completed (FSC1, 2).

2. Survey on existing information on fire failure thresholds and selection of equipment for testing started (FFT1, 2).

3. Experimental work to determine failure thresholds of cables under transient fire environments started (FFT3).

4. Test chambers designed for determining fire failure thresholds of equipments (FFT3).

5. Demonstration of the potential of existing computer code for determining fire environment completed (FED1).

6. Baseline tests designed as basis for extrapolation using computer code (FED3).

7. Final report on electrical cabinet fire testing at University of California - Berkley completed. A second round of testing is planned, and submitted to NRR for review (CR1).

8. Control room systems analysis completed; tests and analysis procedures planned (CR2).

FY 1985 Accomplishements

1. Fire characteristics experiments with representative fuels in different configurations started (FSC3).

2. Fire (including smoke and fire suppressant) failure threshold testing of various selected equipment started (FFT3).

3. Adaptation of computer code to extrapolate fire environments completed (FED2).

4. Baseline tests for determination of fire environment started (FED3).

5. Electrical cabinet fire testing completed (CR1).

6. Control room fire testing started (CR4).

7. Parametric study of control room fire environment by computer simulation started (CR3).

FY 1986 Accomplishements

1. Fire characteristics tests completed (FSC3).

2. Compilation of probable fire characteristics completed (FSC4).

3. Fire failure threshold testing completed and recommendations on testing methodologies for the various equipments prepared (FFT3, 4).

4. Fire environment computer code and user package development completed (FED4).

5. Fire environment baseline testing completed (FED3).

6. Control room testing and computer studies completed (CR3, CR4).

TABLE 1: FIRE PROTECTION RESEARCH PROGRAM
PROJECT TIME SCHEDULE

	FY83	FY84	FY85	FY86
I. Fire Source Characteristic (FSC)				
1. Review of Prior Sandia Fire Tests	─────			
2. Survey of Other Research Results	─────			
3. Fire Source Testing			─────	
4. Summary of Fire Characteristics				─────
II. Fire Environment Determination (FED)				
1. Fire Environment Model Study	───			
2. Computer Code Adaptation for Open Enclosures			───────	
3. Baseline Testing		───────────────		
4. Code Validation and Preparation of User Package				─────
5. Computer Code Adaptation for Compartmented Enclosures				─────
III. Fire Failure Thresholds (FFT)				
1. Establish Operability Requirements	─────			
2. Survey of Existing Information	────────────			
3. Design and Perform Tests		───────────────		
4. Summary of Testing Methodologies and Results				─────
IV. Control Room Habitability and Fire Safety Study (CR)				
1. Testing Electrical Cabinets		───		
2. Control Room Systems Study	───────			
3. Parametric Study of Control Room Fire Environments	───────			
4. Control Room Validation Testing			─────	

Table 2:   FIRE PROTECTION RESEARCH PROGRAM BUDGET ESTIMATES
(amounts in $K)

		FY83	FY84	FY85	FY86
I.	**Fire Source Characteristic (FSC)**				
1.	Review of Prior Sandia Fire Tests	6	35		
2.	Survey of Other Research Results	22	40		
3.	Fire Source Testing		50	100	50
4.	Summary of Fire Characteristics				50
II.	**Fire Environment Determination (FED)**				
1.	Fire Environment Model Study	84			
2.	Computer Code Adaptation for Open Enclosures		200	150	
3.	Baseline Testing		300	300	200
4.	Code Validation and Preparation of User Package				300
5.	Computer Code Adaptation for Compartmented Enclosures				50
III.	**Fire Failure Thresholds (FFT)**				
1.	Establish Operability Requirements	31	80		
2.	Survey of Existing Information	31	50		
3.	Design and Perform Tests		250	250	150
4.	Summary of Testing Methodologies and Results				50
IV.	**Control Room Habitability and Fire Safety Study (CR)**				
1.	Testing Electrical Cabinets	27	100	50	
2.	Control Room Systems Study	7	185		
3.	Parametric Study of Control Room Fire Environments			50	
4.	Control Room Validation Testing			100	100
	TOTALS $K	208	1290	1000	950

APPENDIX A
DETAILED PROJECT DESCRIPTIONS

A.1 Fire Source Characteristics (FSC)

Project FSC1 - Survey of prior Sandia fire tests. Determine the mass and energy release characteristics of the full-scale tests performed by Sandia at Underwriters Laboratories and of other smaller fire experiments involving other fuels.

Project FSC2 - Survey of other existing research results. Assemble the mass and energy release rate characteristics of probable source fuels, such as flammable liquid pools and cable bundles, from the literature. Collect data on the evolution of various corrosive gases.

Project FSC3 - Fire Source Characterization Test. The results of the above projects will be reviewed to identify tests needed to fill gaps in the fire source characteristics. The fuel configurations in such tests will be well defined. The test fires will not be restricted by lack of air.

Project FSC4 - Summary of Fire Characteristics. A data base on the mass and energy release histories of a range of fires will be prepared.

A.2 Fire Environment Determination (FED)

Project FED1 - Fire environment model and computer code selection. Existing fire models and computer codes have been surveyed and selection has been made of one most suitable for development to address the problems of determining fire environments in nuclear power plant enclosures with typical layouts of safety-related equipment. The computer code is capable of describing the environmental parameters as functions of the time elapsed from the inception of the fire until complete suppression. The following requirements were stipulated for the code:

1. The model and the numerical code in its basic form must have already been utilized in enclosure/fire plume studies.

2. The fire source will be simulated by its transient characteristics such as temperature, velocity, energy and mass release rates, and the composition of the combustion products. These and the boundary conditions pertaining to the geometry of the enclosure will be determined independently of the code. The geometrical configurations of the fire source will be defined, and the code will have the capability to incorporate it.

3. The code will have the capability to simulate typical enclosure geometries and to include doors, vents, and obstacles such as equipment, barriers, and beams. It will also have the capability to simulate ventilation conditions, both natural and forced.

4. The numerical procedure of the code will make the error-bounds estimable, and a balance will be struck between the computation time and accuracy.

5. The output of the code will make possible a three-dimensional mapping of the progressively developing environment at any point in the enclosure. The parameters quantifying the environment will, as a minimum, include temperature and velocity.

Project FED2 - Computer Code Adaptation for Open Enclosures. The selected computer code will be adapted to the specific needs of this program in validating the code under CR4 and doing parametric studies of control room environments under CR3 and for control room PRA's under the RMIEP.

Project FED3 - Baseline testing. A series of full-scale fire tests will be performed on a set of generic enclosure configurations. The mass and energy release of a range of source fires obtained from FSC, above, will be reproduced in these tests. The instrumentation will monitor the various thermodynamic variables throughout the enclosures. The tests will cover a range of geometrical variables (length, width, and height ratios) and a range of fire severities, so that the test results can be used to extrapolate to other situations and to validate the computer code.

Project FED4 - Code validation and preparation of user package. The fire environment computer code will be validated by comparison with the results of the baseline test (FED3) and the control room validation tests (CR4). A complete package comprising the necessary software, a full listing of the program statements, flow charts, and user instructions will then be developed.

Project FED5 - Computer Code Adaptation for Compartmented Enclosures. If the computer code is deemed valid for an open enclosure, the code will be adapted for the purpose studying the potential effects of a fire in the compartments of the control room complex on the environment of the control room.

A.3 Fire Failure Thresholds (FFT)

Project FFT1 - Selection of the equipment to be studied. RMIEP will input to this project lists of critical fire areas in the La Salle plant and safe shutdown equipment within those areas. Equipment to be studied for the purpose of determining fire failure thresholds will be selected on the basis of susceptibility to degradation of component materials in a fire or fire suppressant environment. The minimum requirements for operability for each class of selected equipment will be determined.

Project FFT2 - Survey of existing information. Manufacturer's specifications, laboratory research results, and other information on fire failure thresholds of equipment will be collected. Likely sources such as EPRI, EEI, and the national laboratories will be explored.

Project FFT3 - Fire failure threshold tests. Tests will be performed to fill gaps in the knowledge on fire failure thresholds of shutdown and safety-related equipment. The limiting parameters, e.g., temperature, heat flux, corrosive species concentration will be determined. The effect of the duration of exposure, e.g., the time integrals of temperature and heat flux, will also be determined. Tests will expose selected equipment to extinguishing agents, such as water and CO_2, in absence of fire to evaluate failure modes and thresholds.

Project FFT4 - Summary of results and testing methodologies. The testing methodologies for the selected classes of equipment will be described and the data base of fire and fire suppressant failure thresholds for such equipment will be summarized.

A.4 Control Room Habitability and Fire Safety Study (CR)

Project CR1 - Control room cabinet fires. This project will determine the characteristics of fires within control room cabinets and the effects of such fires on the environment in adjacent cabinets. It will develop the characteristics (e.g., heat release rate, smoke generation rate) for several cabinet styles and combustible loadings. This project will determine the range of exposure fires within a cabinet required to fully involve the cabinet contents (cable insulation).

Project CR2 - Control room systems study. A control room systems analysis will be performed, using typical plant configurations, to determine fault propagation through the system in the event of fire and the effects of these faults on the operability of the safe shutdown equipment from the control room. Consideration will be given to the electrical systems aspect of design which limit fault propagation, such as circuit-breaker devices, to the capability to control and monitor shutdown activities from a remote control panel, and to emergency procedures.

Project CR3 - Parametric study of control room fire environments. The development of fire and fire suppressant environments in a generic control room geometry will be studied with the aid of computer simulation. In order to assist RMIEP, the initial study will be plant-specific, using the La Salle plant control room configuration. The energy and combustion products (smoke and toxic gases included) of the fires obtained from the tests of Project CR1 will be used as input parameters to the computer code, as also various rates of ventilation. The simulation will map the room environment parameters, such as temperature and velocity. These maps will provide estimates of smoke and toxic product removal times and periods of vision obscuration--quantities relevant to the habitability of the control room. The study will also indicate the possibility of fire spread to ordinary combustibles remote from the fire

source and of the afterburn of excess pyrolyzates from the fuel in the hot ceiling layer. The findings of this study will be compared to the results of Project CR4,
Control Room Validation Test, for qualitative agreement. The test is on a partial replication only of control room configurations.

Project CR4 - Control Room Validation Tests. Fire testing of a partial replication control room is planned for the purpose of validation of the computer code developed under Project FED2. The source fires will be reproductions of the cabinet fires of Project CR1. The enclosure will be instrumented to monitor the progressively developing environment. The results will supplement the data base developed under Project FED3 for validation of the code, but with emphasis on control room type configurations in this project.

APPENDIX B
COORDINATION WITH NRR/CMEB

B.1 **Fire Source Characteristics (FSC).**

    1.   Survey of prior Sandia fire tests: completed.

    2.   Survey of other existing research results: completed.

    3.   Fire Source Testing

        CMEB to be informed of results of surveys (1,2,3), configurations to
        be tested prior to test performance and subsequent test results.
        Basis for configuration, and test objectives and products should be
        included.

    4.   Summary of Fire Characteristics:

        CMEB will review final draft prior to publication.

B.2 **Fire Environment Determination (FED)**

    1.   Fire Model and Computer Code Selection: completed.

    2.   Computer Code Adaptation: CMEB will review results.

    3.   Baseline testing:

        CMEB will review test plan prior to execution. CMEB will be promptly
        informed of test results. Test configuration, basis for configuration,
        test objectives, and end products should be included with the test
        results.

4. Code validation:

CMEB will be informed of the validation results.

5. Combustion product migration:

If performed, CMEB will be informed of the results.

B.3 <u>Fire Failure Threshold (FFT)</u>

1. Identification of equipment to be tested:

2. Survey of existing information:

3. Fire Failure Threshold Tests:

CMEB will be informed of equipment selected, their operability requirements, and estimated fire failure thresholds.

Prior to testing, CMEB will review the test plan for each equipment type selected. Test configuration, basis for configuration, test objectives, and end products should be included in the test plan.

CMEB will be promptly be informed of the test results.

4. Summary of testing methodologies:

CMEB will review final draft prior to publication.

*U.S. GOVERNMENT PRINTING OFFICE:1985-461-721:20228

NRC FORM 335
(7-84)
NRCM 1102
3201, 3202

U.S. NUCLEAR REGULATORY COMMISSION

BIBLIOGRAPHIC DATA SHEET

SEE INSTRUCTIONS ON THE REVERSE

1. REPORT NUMBER (Assigned by TIDC, add Vol. No., if any)	
NUREG-1148	

2. TITLE AND SUBTITLE	3. LEAVE BLANK
Nuclear Power Plant Fire Protection Research Program	

4. DATE REPORT COMPLETED	
MONTH	YEAR
April	1985

5. AUTHOR(S)	6. DATE REPORT ISSUED
A. Datta	MONTH / YEAR
	July / 1985

7. PERFORMING ORGANIZATION NAME AND MAILING ADDRESS (Include Zip Code)	8. PROJECT/TASK/WORK UNIT NUMBER
Division of Engineering Technology Office of Nuclear Regulatory Research US Nuclear Regulatory Commission Washington, DC 20555	9. FIN OR GRANT NUMBER

10. SPONSORING ORGANIZATION NAME AND MAILING ADDRESS (Include Zip Code)	11a. TYPE OF REPORT
Same as Item 7	
	11b. PERIOD COVERED (Inclusive dates)

12. SUPPLEMENTARY NOTES

Prepared in cooperation with the Office of Nuclear Reactor Regulation, U.S., Nuclear Regulatory Commission.

13. ABSTRACT (200 words or less)

A program plan for nuclear power plant fire protection research has been presented in this report. The principal objective of the program is to create a data base that would reduce the uncertainties in fire probabilistic risk assessment of plants. A three-pronged approach of characterization of potential fires, determination of the ensuing environment, and determination of failure thresholds of safety-related equipment in that environment is described. The techniques are to be applied to estimating the fire safety margin available in a control room.

14. DOCUMENT ANALYSIS — a. KEYWORDS/DESCRIPTORS

Fire energy release; fire environment; safety-related equipment; control room.

b. IDENTIFIERS/OPEN-ENDED TERMS

Fire source characteristics; environment determination; component failure thresholds; control room

15. AVAILABILITY STATEMENT
unlimited

16. SECURITY CLASSIFICATION
(This page)
Unclassified
(This report)
Unclassified

17. NUMBER OF PAGES

18. PRICE